AF694641

# Building A Solar-Heated Pit Greenhouse

**By Greg Stone**

If you've always wanted a greenhouse, but you're scared off by the high cost of energy, take a close look at the solar pit. It's a greenhouse that combines the energy-conserving features of the traditional pit greenhouse with some newer techniques and materials used in modern, heated greenhouses. The result is a simple, inexpensive structure that blends in with almost any environment and requires little or no heat from any source except the sun.

## Heat-Conserving Characteristics

A typical conventional greenhouse is eight feet high and of glass-to-ground construction. This means the glazing material, whatever it may be, extends to just a few inches above the ground. This design provides some light for growing plants under the bench, but it also lets out a significant amount of heat. Heat is lost through the glazing or through tiny cracks between panes of glazing, or through joints in other parts of the structure. In colder climates, greenhouses frequently are built on three-foot sidewalls of concrete block. This decreases heat loss, but is not nearly as energy-efficient as the pit design.

The most distinctive characteristic of the solar pit is that the walls of the greenhouse are underground. They are constructed to ground level of concrete and two inches of foam insulation; sometimes they are simply concrete piers with foam insulation lining the perimeter from below the frost line to ground level.

There are two heat-conserving benefits to this system. First, the walls are in contact with the earth. Since the earth below the frost line remains a pretty consistent 50° F. year round, the walls are significantly warmer than the outside air in winter. In addition, the pit design means the walls are protected from the cooling effects of high winds.

## Heavily Insulated

Not only is heat loss through the walls cut to just about zero, but also the solar pit is heavily insulated. From a heat standpoint, the traditional greenhouse is a big loser because it is walled and roofed entirely in glass, or some other transparent material, with a minimum of framework to block the light. For example, if you are trying to maintain a temperature differential of fifty degrees between the greenhouse and the outside, it takes eighteen times as much heat in an all-glass structure as in a structure entirely insulated with four inches of fiberglass. If you use a double layer of glass — a good conservation measure — heat loss can be reduced as much as 40 percent. At best, however, you are still losing about ten times as much heat through the glass as through an insulated wall.

In the solar pit greenhouse, the pit sides, north-facing wall, and north roof are heavily insulated. The east and west walls are insulated if the greenhouse is large enough to sacrifice the extra light and heat which normally would come through these walls. Only the south-facing roof provides light and heat for plant growth during the day. It is also where most of the heat is lost. To minimize this heat loss, the roof is double glazed, leaving a dead air space between the layers. (Quite frequently this second layer is merely a large sheet of plastic film taped to the inside of structural members.)

*Here's my solar pit. That's the south wall on the right. The north wall is made of asbestos cement board. If I were building it again, I would insulate it with three- or four-inch fiberglass batts. I would add an equal amount of insulation to the lower half of the door.*

## Cut Heat Loss

If effective movable insulation is used over this glazed area at night, the heat loss can be cut in half. In the traditional pit, canvas tarps filled with insulation were employed. This is still a viable alternative, but today people are turning to thermal blankets or insulated shutters two inches thick of polystyrene foam or a similar material. These shutters can be mounted on either the inside or the outside of the greenhouse.

Other energy-saving options include a reflective coating on the north wall to increase light, a reflective curtain on the inside of the south wall that can be pulled down at night to bounce energy back to the greenhouse benches and floor, and insulated shutters that reflect light and heat into the greenhouse during the day.

Finally, some pit designs employ a passive heat storage system if there is sufficient room. The most common technique is to use water in either fifty-five-gallon drums or one-gallon containers. These are lined up against the north wall where they collect heat during the day and release it at night.

## Other Advantages

There is more to this concept than just energy thriftiness, although that's reason enough to choose the pit design. I like the pit for its low profile, and have found it melds into the landscape in an attractive fashion. Its low profile also makes the pit an attractive addition to a house as an attached greenhouse.

The solar pit is a simple structure to build. After all, one-third of it is underground, and in its most complex form, that part is nothing more than a basement. You will probably find it even simpler than this, for no floor is needed. In fact, with my design the only excavation is a trench two feet wide and four feet deep between the raised growing beds. No basement walls, as such, are needed. Instead, the greenhouse rests on pilings to below the frost line. The rest of the construction involves a few basic carpentry skills and only the tools an average homeowner already has.

### OTHER USES FOR THE PIT

If you halt all gardening activities in your pit in late fall, don't let the greenhouse sit there doing nothing. It makes an ideal storage space for late crops such as carrots, beets, and other roots. Store them in sand in boxes, and place the boxes on the floor, shaded from the direct rays of the sun.

Another fine use for the pit greenhouse is in the early spring — on one of the first sunny days. Go out there with a summer chair and bask. It's certain to make you realize that summer can't be too far away.

# Design

The construction details of your solar pit will depend on how you plan to use the greenhouse. A pit greenhouse can be used in three different ways. The first is as a traditional, heated, indoor garden for growing all sorts of crops. Under the worst conditions, solar energy should keep the solar pit at a nighttime minimum of about 40° F. To maintain an extra ten or fifteen degrees (50°-55° F.) opens the greenhouse to just about every imaginable crop from orchids to tomatoes. You can do this at little cost and with a clear energy conscience, for you will be using far less heat from artificial sources than you would with a greenhouse of traditional design.

A greenhouse can be used primarily for cool-weather vegetables, flowers, and herbs, at least during the winter. It doesn't require any supplemental heat. However, chances are that greenhouse temperatures will drop to near freezing some nights and, depending on the climate and construction, could even drop below freezing. This means the greenhouse can be used only for hardy crops, such as marigolds, lettuce, and radishes — at least in mid-winter. (Some crops can stand brief spells of freezing temperatures.)

The third use is as a season extender — for starting young plants a couple of months ahead of time in the spring and for keeping flowers and vegetables thriving well past the first frost in the fall. This eliminates the greenhouse from any gardening plans during the worst of the winter weather — from about the first of December until mid-February in northern states.

The main design consideration for these different uses is whether or not to insulate the end walls and the north roof. The end walls can contribute significant amounts of light in a small greenhouse, especially in the spring and fall. If your purpose is just as a season extender, you may want to put up with the extra heat loss (after all, the greenhouse won't be used during the coldest weather) in order to get more light, especially for young plants in the springtime.

Consider also the various light and temperature needs of any special crops you may have in mind. For example, if you are a cactus collector, it may be better to sacrifice some heat to light by leaving the end walls uninsulated. Cacti can stand temperatures that drop near freezing occasionally (in fact, some can take temperatures below freezing), but they do like plenty of light. On the other hand, an orchid or African violet fancier could easily forego some light in favor of heat conservation.

## Make It Flexible

The ideal solution is to design a flexible greenhouse which can be adapted to all of these functions. My initial pit design used fiberglass insulation only in the north wall. Even the north roof was transparent. However, I had three-inch styrofoam blocks which clipped in place and were covered with a sheet of polyethylene. This way I could insulate the north roof during the three coldest months, yet leave it clear in the spring for optimum light for seedling growth. Such movable insulation could be applied to the end walls, either in the form of insulating shutters or a more semipermanent arrangement such as the one I use on the north roof of my pit.

Let's take a closer look at the various design and construction details.

## The Roof

The slant of the south-facing roof is a major design consideration. Traditionally both the north and south roofs have been at a 45° slant. This also is an easy angle to work with in construction.

Recent tests show the traditional thinking isn't bad, for the optimum roof angle is only at its best for one day of the year. The reason is that this angle varies with the angle of the sun, and that changes continuously. The optimum angle for the south-facing roof of a solar pit that will be used year round can be arrived at by the rough rule of thumb of adding 20 degrees to your latitude. However, there are so many variables that when members of a research team at Brace Research Institute in Canada examined the various possibilities on a computer, they concluded that any slope from 40° to 70° would be effective. Because of the specific materials available, they chose a slope of 35°, deciding that this wouldn't hurt the solar performance of their structure that much.

Generally speaking, the higher angles for the south roof will do better in mid-winter. Lower angles, closer to 45°, mean your greenhouse will perform best in spring and fall, so again, go back and take a close look at the intended use.

Consider the angle of the north roof. To get the best reflection of light, it should be slanted between 60° and 75°. However, the slope of my north wall is much gentler because I built a two-foot high north wall above-ground to create more vertical growing room in the north "bench" area. The wall also provides more head room and space for water containers, should you want to store excess heat.

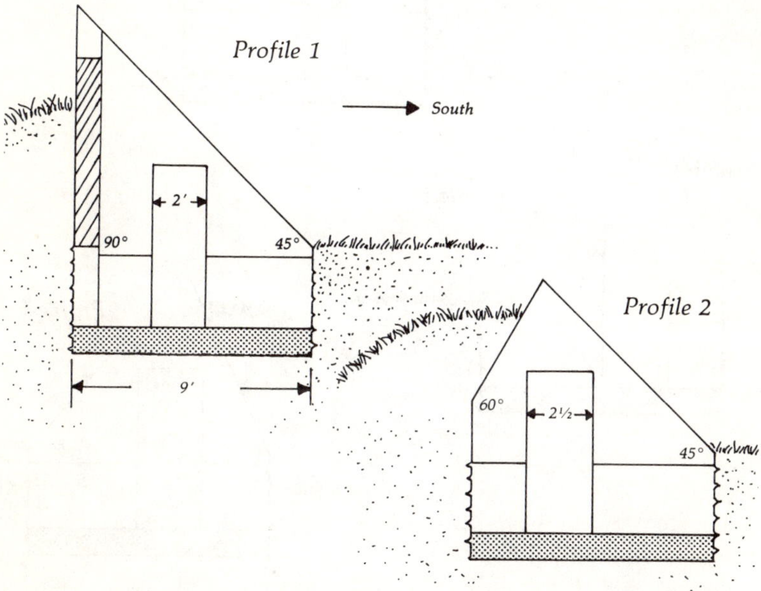


*Here are several good configurations for a solar pit, and there are many more. The basic idea is to have a south-facing roof with an angle of between 40 and 60 degrees.*

*Profile 1 uses a nine-foot roof and provides lots of water storage. This cuts into the interior space, and to keep a reasonably wide north bench, the door has been reduced from thirty inches wide to twenty-four inches.*

*Profile 2 is a compromise providing some wall space for heat storage, as well as a roof slanted for good reflection. (Good reflection will be achieved with the north roof-wall slanted at sixty to seventy-five degrees.*

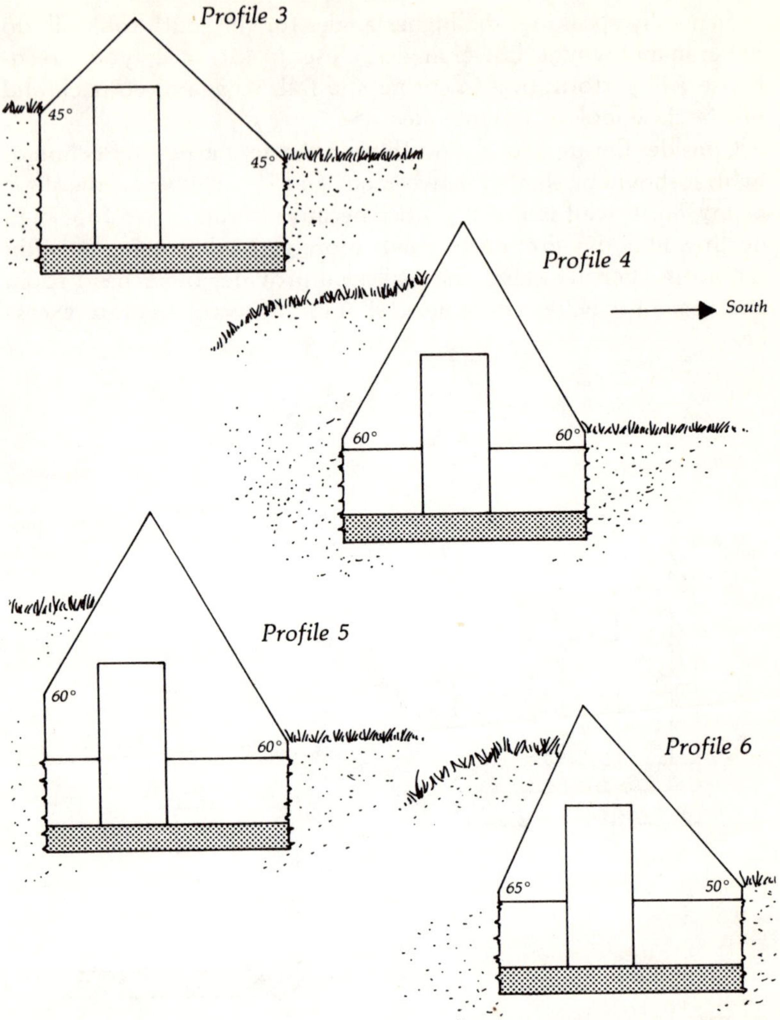


*Profile 3 is a compromise in favor of keeping a low profile for aesthetic reasons.*

*Profile 4 uses sixty-degree slants on both roofs. This minimizes growing space, especially vertically above the north bench and as with the other designs that have high peaks, there's a lot of area near the roof that will be heated, but won't be of much use in growing plants.*

*Profile 5 is similar, but offers more growing height over the north bench.*

*Profile 6 offers an idea of how the angles can be varied within the basic ranges given to come up with other combinations that should work, but may be more pleasing or practical under certain building conditions. In all cases the excavated earth is used as a berm to the north.*

## Location

Before choosing a design, look for a site that has good drainage; is protected by trees, buildings, or whatever, from the worst of the north winds; and will get a full dose of winter sun.

The lack of good drainage can doom a pit greenhouse. Check the water table before you get started, and remember it changes from season to season. If drainage may be a problem, install drainage tiles along the footings with the drain sloped away from the foundation. Cover the tiles with six to eight inches of coarse gravel, and put plenty of small rocks in the bottom of the pit. If you build a foundation wall of concrete block, the foundation can be sealed with cement plaster or asphalt waterproofing. You'll find most land either is naturally suited to a pit, or can be adapted, but some locations are simply too wet.

Try to avoid frost pockets, such as you'll find at the bottom of a hill. On the other hand, a hill to the north, or at least a good stand of shrubs or evergreens, can act as a windbreak and save some heat. You can of course alter the landscape in your favor by taking the dirt that is excavated for the pit and heaping it on the north side of the structure as an earth berm; you can get further protection by planting a windbreak on this side of the pit. However, don't expect miracles. It takes about two feet of dry earth to equal the insulating qualities of just one inch of fiberglass.

Check the trees to the south of you. An evergreen can cast a very long shadow, even at noon, when the sun is low on the horizon in winter. It's amazing how many gardeners forget that trees grow. A friendly-looking, five-foot hemlock may be no threat now, but wait until it reaches thirty feet! Suddenly it's cutting off light at the most crucial time of year. A deciduous tree, however, might be a boon, providing shade in summer, yet blocking little sun in the winter. Look for these things in advance, then you won't discover that some year you must choose between the greenhouse and a tree.

## Attached or Detached

Greenhouses can be attached to your home or free-standing structures. Attached greenhouses have several advantages. They are easier and cheaper to build; there is one less wall, and hook-ups for water and electricity are simpler to install. The attached building serves as a warm north wall with good protection from the wind. In addition, it is far pleasanter to step from one room of your home into an attached greenhouse in winter than to put on boots and a coat and trudge through the snow to get to a free-standing one.

**GRADUAL CHANGE**

We do not want to raise false hopes or predict miracles. We only know that plants hardened off gradually in the open garden or in an unheated greenhouse will tolerate lower and lower temperatures even if the mercury drops below freezing. Each plant has a capacity to adapt itself to changing conditions. If the change is sudden and drastic, the plant may be destroyed. But if the change is gradual, allowing the plant to adapt itself to the alterations, there is no doubt in our minds concerning the ability of many plants to make the adjustment needed for their survival.

Helen and Scott Nearing
*Building and Using Our Sun-Heated Greenhouse*
Garden Way Publishing $6.95

## Use the Heat

You may be able to establish a symbiotic energy relationship between the attached buildings. Excess greenhouse heat, and there will be plenty on a sunny day even in the middle of winter, is vented into the house to help warm it; then at night heat from the house is ducted back into the greenhouse.

Some people choose the detached solar pit because they have no convenient way to attach a greenhouse to their home. If the pit of an attached greenhouse is to have direct entry into the main building, that entry probably will be into a basement, not usually the most convenient spot. Also you will have to deal with a foundation wall. Check with an engineer or professional builder to make sure there is no problem with digging a four-foot deep pit along the foundation of your home. (You may even find that a pit greenhouse can be designed to squeeze underneath a first-story window, thus making it possible to attach a pit greenhouse where in some instances a normal lean-to model would present problems.)

There are other reasons to choose a detached greenhouse. Your home may not have a southern exposure that can be adapted to this use, and an exposed south-facing wall is essential for a solar greenhouse. It also may be more convenient to work a detached greenhouse into a total landscape plan that considers the proximity of garden shed, compost piles, orchard, and vegetable garden. Others choose the free-standing greenhouse because they are fascinated by the challenge of making an entirely solar-heated structure.

# Size

The size of your greenhouse probably will be limited by your ambition and your pocketbook, but the best advice is to build the largest greenhouse you can afford. Small greenhouses may look cute, but they are energy-inefficient, uneconomical to build, and present their own special problems.

A small greenhouse tends to heat up quickly and cool off quickly, making for an unstable environment, hardly ideal for plants. This occurs because the ratio between the volume of air in a small greenhouse and the surface area of walls and roof through which heat may be lost is very high.

Generally the small greenhouse will cost more per square foot. Items purchased in large quantities tend to be less expensive. In addition, costs for having electricity and plumbing installed, or for putting in a ventilation fan will be the same for a fairly wide range of sizes.

My 8′ × 12′ pit is about the minimum size, and even with it the east and west walls are left transparent. If they had been insulated, they would block too great a percentage of sunlight. In larger greenhouses — say twenty feet long — the east and west walls can be insulated and this will save considerable heat. Larger greenhouses are more economical, and most gardeners find they can fill to overflowing almost any size greenhouse in a surprisingly short time.

Before making the final determination of the dimensions of your greenhouse, consider two very practical questions:

1. How far can you reach? For most of us, even those raised in the most demanding of boarding houses, it is inconvenient to reach much beyond three feet. Thus, if you designed a greenhouse with a floor plan that required a four-foot stretch to work the extreme edges of the beds, most of that last foot of space would be wasted unless used for heat-storage water containers.

2. What are the dimensions of the materials you will use in your construction? Sheets of whatever glazing material you select are sold in specific lengths and widths. Know those widths and lengths before deciding on your pit dimensions, so you don't waste valuable time and material cutting those sheets to fit the dimensions you select.

## Foundation

Once you have decided on the size and general location of your greenhouse, you are ready to begin excavation.

Locate the corners of the greenhouse and mark them with stakes. To insure that the lines form a rectangle, use the 3-4-5 principle (Pythagorean theorem). Measure three feet from a corner along one edge and mark it. From the same corner measure along the other edge four feet. The distance between these two points should be five feet. You may have to adjust the corner stakes a couple of times before you find the exact locations. If you have measured accurately, the diagonals should be equal in length.

Next set up batter boards four feet away from the corners. Stretch strings between the boards. Where the strings intersect should mark the corners of the foundation wall. A plumb bob will help in establishing these points. Use a straightedge and a four-foot carpenter's level, or a line level, to check that the strings are level. During diggings the strings are removed but are readily available for reference.

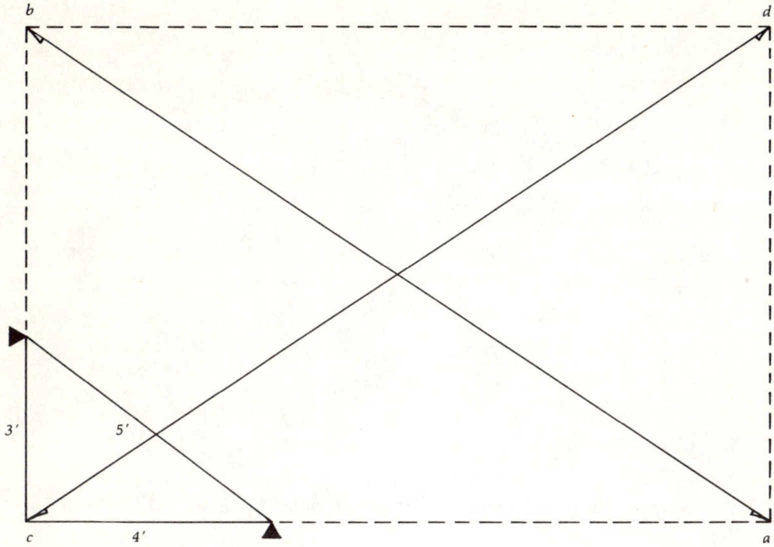

*To get a rectangle with right angles use this method. Line ab and line cd should be the same length.*

## Digging In

In my pit greenhouse I tried for economy, both of labor and money, and so I eliminated much of the pit and foundation work that you will find in other designs. I dug a 16-inch ditch around the perimeter. In that ditch I used a post hole digger to dig six-inch post holes down to below the frost line. These were on three-foot centers. I poured concrete into the six-inch holes, then, when the concrete had hardened, I completed the piers by placing a concrete block on top of the concrete posts. The boards of insulation were placed between the posts, and the dirt shoveled back around them.

A warning: take a lot of care in making certain the piers are level with one another. The height of the pier can be adjusted a bit by varying the thickness of the mortar joints.

When ready to put the anchor bolts in place, fill the holes in the concrete blocks with concrete, let it begin to set, then put the anchor bolts in place, leaving enough protruding so that they will extend through the sill, and a nut can be placed on them.

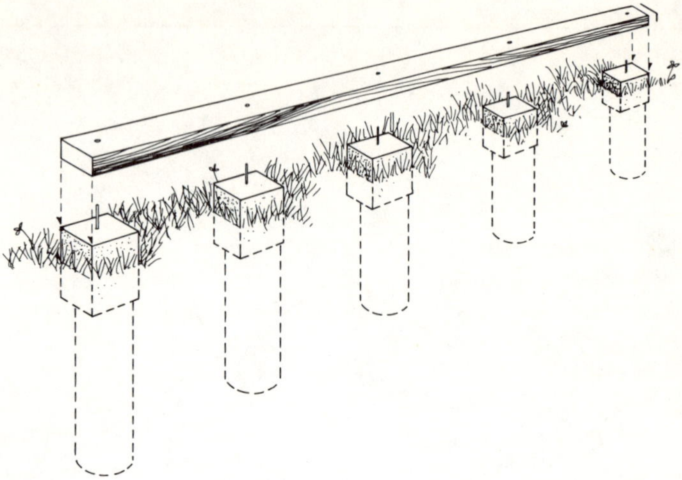

*It's necessary to have all piers at the same level if the 2 × 6 sill is to fit correctly.*

Into the top of each pier cast an anchor bolt of 1/2″ × 8″. Make sure it is positioned correctly. Drill holes in the 2 × 6 sill for the anchor bolts and attach the sill to the piers. The rest of the greenhouse will be attached to the sill.

Instead of digging a full-sized pit, I dug a center aisle, two feet wide and four feet deep which ran the length of the greenhouse minus 3 feet for the bench opposite the door. In the bottom of the pit went a one-foot layer of gravel. The sides of the trench were shored up with a framework of 2 × 4's sheathed with 1/8-inch asbestos cement board. The excavated dirt was later used to make an insulating bank against the north wall.

## Soil Benches

The rest of the soil surface became my "benches." This soil I turned over to a depth of about two feet and enriched with liberal amounts of compost and manure. Take care to make sure this soil has excellent drainage and fertility, for it will receive intensive use over the years. Most of my crops are planted directly into the soil; however some are planted in pots placed on top of the soil, and seeds are usually started in separate flats using a sterile medium to avoid damping-off. Follow the initial preparation with regular supplements of manure and compost to keep the soil rich and well-structured.

This method of constructing a solar pit saves labor; it eliminates the need to build benches, as well as the need to heat the space beneath them. Although earth isn't as good at holding heat as water, it does provide a little extra heat storage. Finally, this design eliminates the need for a foundation.

If you plan a footing foundation of concrete blocks or poured concrete, be sure to include a footing that is as deep as the wall it supports is thick, and about twice as wide. *The footing must be below the frost line.* Greenhouses are viewed as relatively light structures and so are frequently seen as needing a minimal foundation. But that's the traditional greenhouse. The solar pit is heavier with its insulated walls and roofs, so make sure you build a suitable foundation. If in doubt, check with a professional contractor, architect, or engineer. Also, see your local building inspector to insure you are following any applicable codes.

Whether piers or a full-sized foundation is installed, insulation should be included to *below* the frost line. Use two inches of styrofoam boards around the perimeter. (If a foundation is used, you can get by with one inch of foam. If no foundation wall is used, screen the foam with half-inch hardware cloth to discourage pests from burrowing through.) The in-ground insulation will go a long way toward keeping heat in the greenhouse and keeping the soil temperature up.

## Lot of Work

Don't minimize the amount of work this involves. If you are placing the styrofoam outside the foundation, you've already done all the digging in preparation for that foundation, and the styrofoam simply goes outside the concrete blocks.

If you're digging to place that insulation between piers, and you have to go down three feet or so to get below the frost line — that's a lot of digging, particularly if you aren't doing such hard work each day.

But you can do it if you're healthy and have no back problems. I did the foundation work on mine in a day.

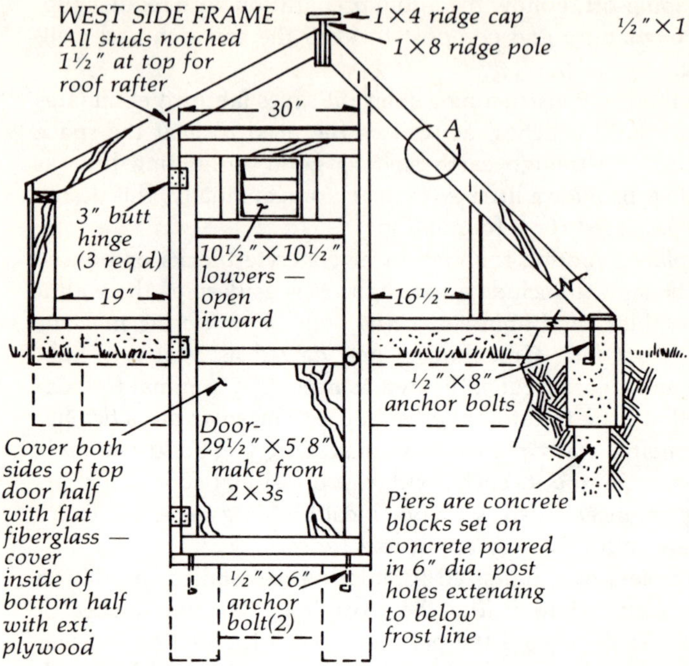
WEST SIDE FRAME
All studs notched 1½″ at top for roof rafter
1×4 ridge cap
1×8 ridge pole
30″
A
3″ butt hinge (3 req'd)
10½″×10½″ louvers — open inward
19″
16½″
½″×8″ anchor bolts
Cover both sides of top door half with flat fiberglass — cover inside of bottom half with ext. plywood
Door- 29½″×5′8″ make from 2×3s
Piers are concrete blocks set on concrete poured in 6″ dia. post holes extending to below frost line
½″×6″ anchor bolt(2)

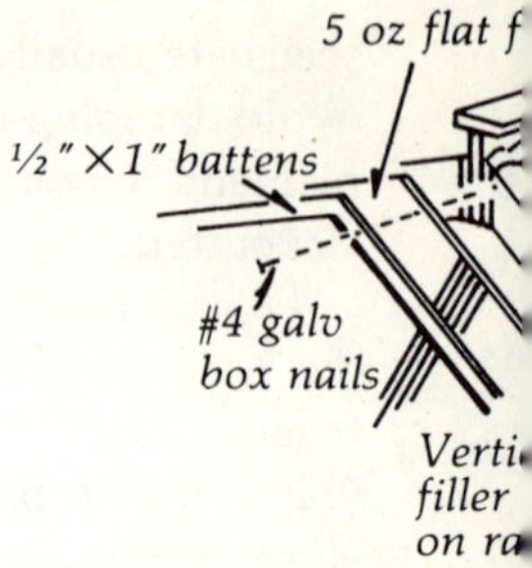
5 oz flat f
½″×1″ battens
#4 galv box nails
Verti
filler
on ra

DETAIL A
Corrug f
Space
Fla
South roof

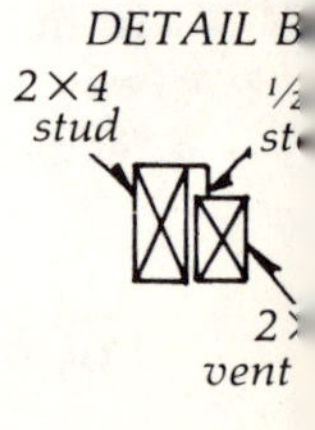
DETAIL B
2×4 stud
½
st
2
vent

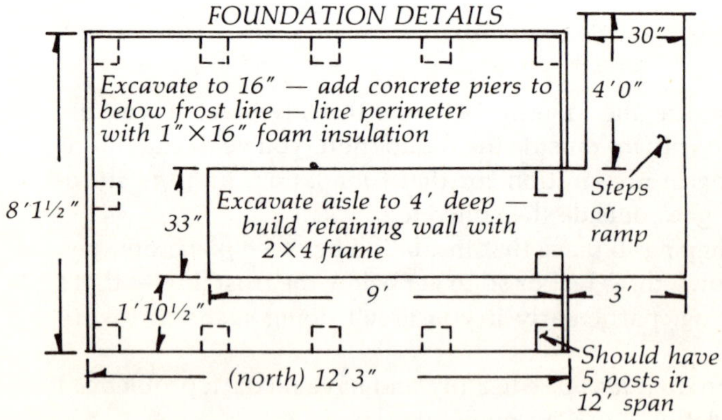
FOUNDATION DETAILS
Excavate to 16″ — add concrete piers to below frost line — line perimeter with 1″×16″ foam insulation
30″
4′0″
Steps or ramp
8′1½″
33″
Excavate aisle to 4′ deep — build retaining wall with 2×4 frame
9′
3′
1′10½″
(north) 12′3″
Should have 5 posts in 12′ span

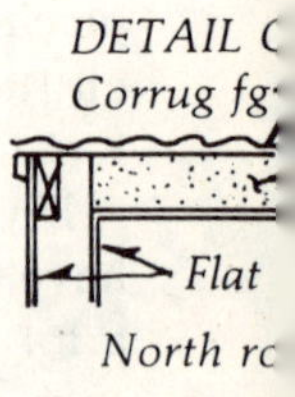
DETAIL C
Corrug fg
Flat
North ro

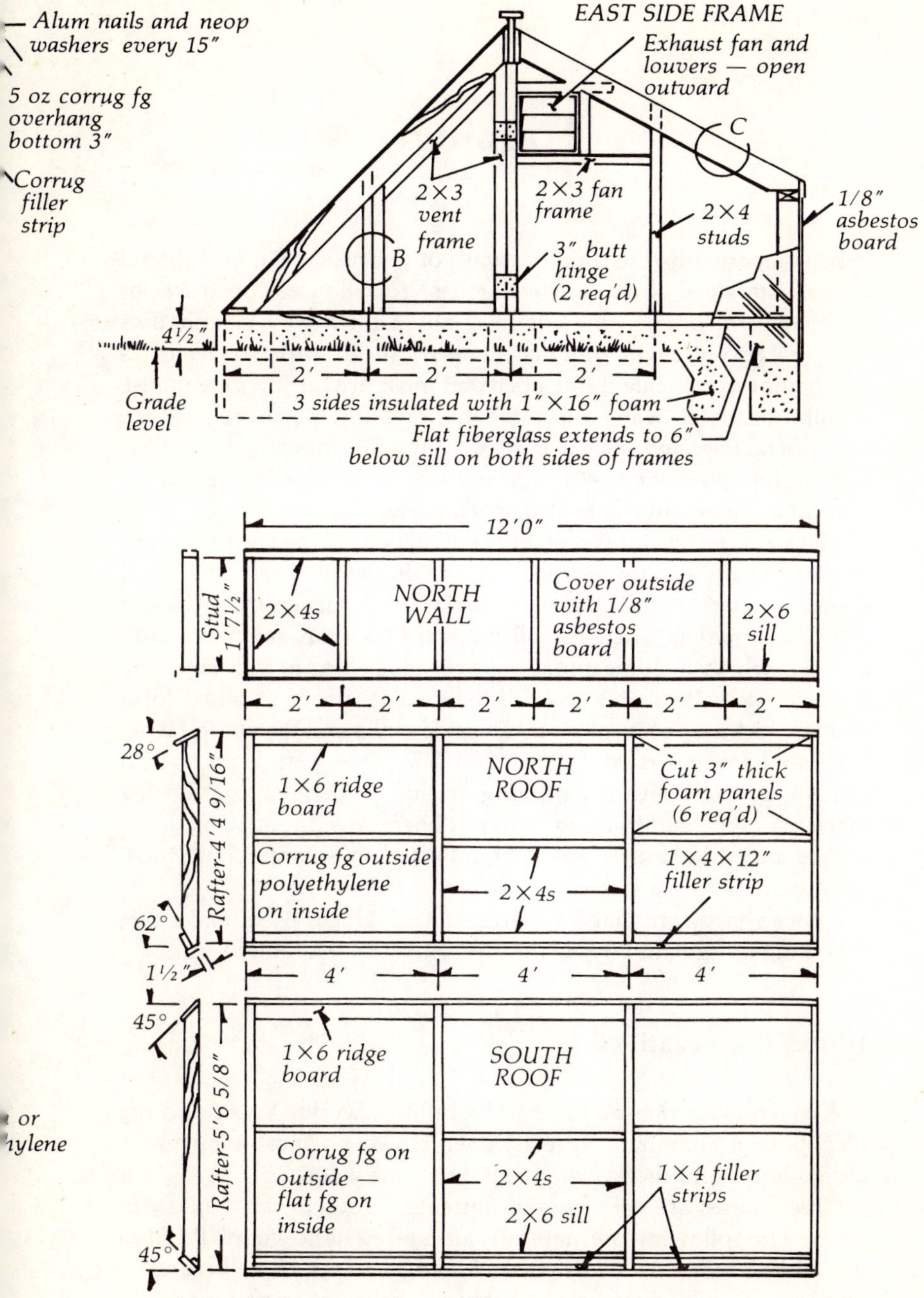
Alum nails and neop washers every 15″
5 oz corrug fg overhang bottom 3″
Corrug filler strip
EAST SIDE FRAME
Exhaust fan and louvers — open outward
C
2×3 vent frame
2×3 fan frame
2×4 studs
1/8″ asbestos board
B
3″ butt hinge (2 req'd)
4½″
Grade level
2′
2′
2′
3 sides insulated with 1″×16″ foam
Flat fiberglass extends to 6″ below sill on both sides of frames
12′0″
Stud 1′7½″
2×4s
NORTH WALL
Cover outside with 1/8″ asbestos board
2×6 sill
2′
2′
2′
2′
2′
2′
28°
Rafter-4′4 9/16″
1×6 ridge board
NORTH ROOF
Cut 3″ thick foam panels (6 req'd)
Corrug fg outside polyethylene on inside
2×4s
1×4×12″ filler strip
62°
1½″
4′
4′
4′
45°
Rafter-5′6 5/8″
1×6 ridge board
SOUTH ROOF
or
ylene
Corrug fg on outside — flat fg on inside
2×4s
1×4 filler strips
2×6 sill
45°

# Framing

When constructing the roof and walls of your solar pit, you should follow standard framing techniques for house construction. Although you'll see some commercially built hobby greenhouses using relatively small framing members in an effort to let in as much light as possible, remember that most of your pit greenhouse is solid, insulated walls. Since the building may not have a source of artificial heat and since heat loss is minimal, snow loads are likely to build up on the north roof and may stay there longer than it will on a conventional, heated greenhouse.

The sizes of structural members will depend upon such things as spans, and the spacing between rafters on any glazed surface will depend in part on the glazing material used. Generally speaking, 2 × 4s should be used for all rafters. (You can get away with 2 × 3s, which in some situations will block less light.) You may want to use rafters of 2 × 6s in the insulated roof to allow for a greater thickness of insulation. The side walls and the south-facing roof can be of standard construction using 2 × 4 studs on 16-inch centers. *Before* doing any framing, decide on the glazing. Glazing materials come in different standard dimensions; the spacing of the rafters and studs may have to be adjusted to accommodate these measurements.

Choose hardware that is rust-resistant. If glue is used, make sure it is a highly moisture-resistant type, such as resorcinol resin.

## Wood Preservatives

A greenhouse doesn't have to be built of wood. You could use PVC piping, aluminum, or even steel. However, most people who design and build their own greenhouses use wood.

Greenhouses are very humid; untreated wood decays rapidly in them. The following are naturally decay-resistant woods listed in descending order: 1. black locust, 2. osage-orange, 3. red cedar, 4. redwood, 5. white oak, and 6. cypress. Some of these woods are becoming scarce and are expensive. Others, such as redwood, are not as strong as some other common types. For these reasons, many people are turning to pressure-treated wood.

Pressure treatment is a process whereby preservative chemicals are forced throughout the wood. The result is a product that will last longer than most of the woods just listed. Pressure treatment does vary, however, so make sure you understand what you are purchasing and whether or not the wood has been treated for in-ground or above-ground use. Also check on the chemicals used in the process. Stay away from oil-borne preservatives, such as creosote. Their fumes are highly toxic to plant growth. Look instead for the water-borne, salt-type preservatives, as most of these are non-toxic to plants. Examples of these are acid copper chromate, chromated zinc chloride, chromated copper arsenate, and fluor chrome arsenate phenol.

# Glazings

Since the solar pit greenhouse has far less transparent surface than a conventional design, experimenting with new glazing materials is not such a costly process. This is good, because while the heat retention characteristics of these new materials looks good, the life-span of some of them is questionable.

## Glass

For years glass was the only choice, and it is still very popular. It does a good job of letting in sunlight and trapping infrared radiation. It also weathers well and should last indefinitely. For greenhouses it is usual to use glass in 1/8-inch thick panes that are 20 or 24 inches wide.

If you decide to use glass, adjust the width between the structural supports to fit the panes. For example, glass panes 20 inches wide will fit between studs spaced 22 inches on center. Stops will be needed to keep the glass in place, and a mastic sealant should be used to decrease air leaks. It may be easier to cut 1/2-inch rabbets in the studs for the glass, in which case the studs should be spaced on 21-inch centers.

There are two problems with glass, however, and they are of special interest to the solar pit builder. First, glass is brittle and shatters easily. This is a safety hazard for the people inside as well as outside the greenhouse. Since the glazing on a solar pit will go to the ground, or to just inches above it, the chances of someone tripping and falling through the glass are real, if remote. Second, while the glass itself is good at trapping heat, the fact that several panes must be mounted on a slanting surface means that there are many opportunities for infiltration and heat loss around the edges of each pane.

## Fiberglass

The most popular alternative seems to be fiberglass. The cost of long-life fiberglass is comparable with glass. Its light transmission and ability to hold in heat (infrared radiation) are also good. In fact, it gives a more diffused light than glass, which is desirable for growing plants.

Generally fiberglass comes in panels that are 26 inches wide and anywhere from 8 to more than 20 feet long. It also is available in rolls 4 or 5 feet wide and 50 feet long. (The panels are usually the corrugated type, while the rolls are flat.) Fiberglass is easy to cut with a saber saw or metal shears and can be mounted with pop rivets, nails, or screws. (A neoprene washer should be used between the head of the nail or screw and the fiberglass. You can buy nails with the washers already on them just for this use.)

For the corrugated fiberglass, framing should be on 2-foot centers, and the fiberglass panels overlapped one corrugation. Apply a clear sealant between the panels to insure a tight joint.

The flat fiberglass needs support only around the perimeter when it is used vertically. If it is used for the roof, space the joists on 2-foot centers and add blocking between the rafters every three to four feet. The flat fiberglass is flexible and will sag otherwise.

The corrugated fiberglass usually is recommended for greenhouse roofs because it is stronger, while the flat fiberglass is suggested for the sides. However, the corrugations add about 10 percent more surface area than a flat sheet covering the same space, and more surface means more area through which heat is lost. For

*Pop rivets can also be used to hold fiberglass in place.*

this reason, many people building solar greenhouses opt for the flat sheets.

Fiberglass won't shatter, even when struck by a baseball or rock, although it can tear if hit hard enough. However, it will burn and this should be taken into consideration, especially if you are constructing an attached greenhouse. You should be extra careful with heaters near fiberglass or any of the plastic films or sheets.

You also should be careful about the fiberglass you purchase. Not only is there a difference between brands, but many manufacturers make several different grades. Do not buy the first available panels at a local hardware store. Chances are they're not designed for greenhouse use and will deteriorate quickly.

The sun is the major enemy of fiberglass. Ultraviolet rays damage it, and while a good panel may give 20 years of service in northern states, it may give far less in the southwest. Check guarantees carefully; make sure the panel is designed for greenhouse use; and look for panels with special coatings, such as du-Pont's *Tedlar*, which screen out most of the harmful rays and protect the panel.

## Other Alternatives

Other alternatives are acrylics, such as *Plexiglass*, and polycarbonate sheets, such as General Electric's *Lexan*. These have fine light transmission and heat retention characteristics, come in large sheets, are resistant to impact, are strong and have a long life expectancy with relatively little deterioration from ultraviolet radiation. Their major drawback is that they cost twice as much (sometimes even more) than glass or fiberglass. They also tend to scratch more easily and present some special mounting problems because they expand and contract with temperature changes. Because of this they should be mounted in a channel with some sort of flexible gasket to make a tight seal.

All of these materials mentioned so far are either rigid, or semi-rigid and relatively thick. There is also a wide variety of thin films applicable to greenhouses, including polyethylene, vinyl, *Mylar*, *Tedlar*, *Teflon* and acetate. Of these, the best known and most frequently used is polyethylene.

*A sealant between fiberglass panels insures a tight joint.*

**ENERGY CONSUMPTION**

A conventional single-glazed greenhouse in southern New England, heated to 60° F. on winter nights probably consumes around sixty-five watts (200 Btus) of energy every hour for every square foot of ground it covers. At seven cents per kilowatt, this works out at an average of about forty-five cents per hour for every one hundred square feet of greenhouse space, which would give a winter heating bill of about $600.

Peter Clegg and Derry Watkins
*The Complete Greenhouse Book*
Garden Way Publishing $7.95

## Polyethylene

Many hobby and commercial greenhouses use a 4-mil or 6-mil polyethylene film because it is inexpensive, requires a minimum of framing, and is flexible enough so that greenhouses need no foundation. (In some communities this means the structure is classified as a "temporary" building and therefore isn't taxed — another reason for going the polyethylene route.)

However, polyethylene may last as little as six months and even the best, ultraviolet-resistant type, Monsanto 602, may last just two or three years. In the long run polyethylene may not be a money saver, although it will keep the initial investment low.

Polyethylene is flammable; it can rip; and it allows most of the heat radiation from inside the greenhouse (infrared) to pass right through. Its main advantage is that it comes in very large sheets, is relatively easy to work with, and can be secured with very few joints to make an extremely tight house.

## Other Films

Vinyl films are a bit heavier, more costly, and also deteriorate in the sunlight, although not as quickly as polyethylene. *Mylar,*

*Teflon* and *Tedlar* have better weatherability and stand up to the sun relatively well (you can expect five to ten years from them), but the cost is significantly higher than for polyethylene.

The film that some solar enthusiasts are turning to is acetate. It does a good job of letting light through and retaining heat. Its poor weatherability makes it a bad choice for outside use, but it is frequently used as the second layer of glazing on the inside of a solar greenhouse. Its ability to retain infrared radiation makes it a better choice for this task, some feel, than polyethylene, although its brittleness tends to make it more difficult to work with. One method for mounting acetate is to glue it to a framework of thin wood, then attach the frame to the large structural members of the greenhouse.

**A COMPROMISE**

*If* you don't provide heat for your pit greenhouse, and *if* you want to make certain the temperature doesn't drop below freezing, there's a compromise you can make.

Run an all-weather extension cord from the house to your greenhouse, link it with a thermostat set for 34-36 degrees Fahrenheit, and hook on two or three 150-watt heat lamps, set so they don't shine directly on any nearby plants. You'll be surprised how infrequently the thermostat calls for power, even during the coldest periods, if you've insulated your greenhouse well.

## Double Glazing

Whatever glazing you choose, you definitely will want two layers with a dead air space in between them. (A dead air space anywhere from 3/4 inch to 4 inches will be roughly equal in effectiveness.) Studies have shown that this second layer reduces heat loss 25 to 40 percent and cuts down about 10 percent of the light — an acceptable trade-off. A third layer, except in extremely cold regions such as Canada, isn't a good trade-off because it cuts out too much light for a relatively small savings in heat. Polyethylene films and similar materials that degrade quickly when exposed to the full effects of sun and weather will last longer when used as the inside layer in a double-layered glazing system.

Finally, there are several glazings now on the market that are built in two layers. These include *Sunwall* made by Kalwall Corporation, 111 Candia Road, Manchester, NH 03103; *Tuffak-Twinwal* made by Rohm and Haas, Independence Mall, Philadelphia, PA 19105; and *Acrylite* made by Cyro Industries, 697 Rt. 46, Clifton, NJ 07015. These glazings have the advantage of a built-in, double-layered system, although the dead air space may be smaller than considered optimum. Before working your design through, you may wish to write each company to get the full specifications on their material. *Sunwall* is a fiberglass material, *Tuffak-Twinwal* is a polycarbonate sheet, while *Acrylite* comes in both an acrylic version and a polycarbonate one.

# The Walls/Roofs

The north roof and the two-foot, above-ground north wall should be heavily insulated with fiberglass batts. If the greenhouse is larger than the minimum size (8′ × 12′), the east and west end walls also should be insulated.

Inside the greenhouse you may wish to paint the north-facing roof flat white (which gives diffuse reflection) to make it highly reflective, or to cover it with a shiny surface, such as aluminum foil (bright reflectors tend to cause "hot spots," but they keep more of the light in the greenhouse than the diffuse reflectors will).

Commercial reflective films, which can be arranged on a system of wires and pulleys, are available from Duracote Corporation, 350 North Diamond St., P.O. Box 512, Ravena, OH 44266, and from Dupont de Nemours and Co., Inc., Textile Division, Wilmington, DE 19898. David MacKinnon used parachute material with aluminum foil glued to one side for a curtain in a greenhouse constructed in Arizona. Appropriate Technology Corporation, P.O. Box 975, Brattleboro, VT 05301 makes a tight-fitting window shade material with an "R" value of 4.75 that might be adapted as a greenhouse curtain.

The south-facing roof should be slanted to make the most of the weak, low-angle sunlight during the cold months. It should be double glazed. We suggest that outer glazing be a flat fiberglass sheet designed for greenhouse use with a life expectancy of at least 15 years. The inner glazing will be polyethylene or acetate with a 3½-inch (the width of a 2 × 4) dead air space between the two. A thermal blanket or shutters should be pulled over this roof at night.

## Shutters

Shutters can be used either on the inside or on the outside of the greenhouse. Some people favor *internal* shutters. A flexible, insulating cover can be made by enclosing fiberglass batts in polyethylene and taping the edges to keep out moisture. This can be clipped on at night and taken off and stored, preferably out of direct sunlight to preserve the polyethylene, in the day. Others find a rigid shutter easier to handle. In this case design your south wall with a 2¼-inch high rim of wood all around the perimeter. Then sandwich 2-inch styrofoam panels between 1/8-inch sheets of temperboard, or some similar material, and clip the shutters in place at night. The problems with such shutters is the need to store them during the day, as well as putting them on and taking them off regularly. I find them too cumbersome in a small pit where plants are likely to be growing close to, or even touching, the glazing material. However, if such shutters are constructed so there is a half-inch dead air space between them and the glazing, heat loss can be reduced by as much as 90 percent.

For my 8′ × 12′ greenhouse I made three 4′ × 5′ insulated *exterior* shutters which protect the south wall at night but can be removed each day for storage. (The number and size of the shutters will vary depending on the pitch and length of your south roof.) To the sides and bottom of the greenhouse roof, I added 1 × 3s to hold the shutters. A new cap may be necessary across the top of the greenhouse. Make the shutters of at least one-inch thick foam which is covered with 1/4-inch exterior plywood, canvas or any other opaque material that will reinforce the foam and protect it from the weather. Daytime storage may be a problem, but if the earth berm is sloped, the shutters can be shifted to the north roof during the day.

# Heat Storage

For years pit greenhouses have gotten by with no special attempts at storing excess heat, but useful gains can be made with a good water or rock storage system. Water is the more efficient storage material because of its capacity to hold heat.

A water storage system usually consists of bottles with dyed water, or black, 55-gallon oil drums filled with water, arranged against the back (north wall) of the greenhouse. The current trend seems to be away from large containers in favor of small, 1-5 gallon containers stacked against a north wall, usually on shelves. You might want to write One Design, Inc., Mountain Falls Route, Winchester VA 22601, for information about their modular water-wall system.

## Take Up Much Space

Such water storage systems are in direct competition for sunlight with the plants. They take up good growing space and give you a black north wall — black absorbs heat best.

Rock storage solves some of these problems. Fist-sized rocks, in a bin either under the floor or behind the north wall, store the excess hot air generated in the solar pit on almost any sunny day, even in the middle of winter. A fan blows this hot air over the cooler rocks which absorb the heat during the day; at night the fan draws off the warm air to heat the cooler greenhouse air. This method won't store as much heat as a smaller volume of water, but it does have the virtue of *not* being in competition with the plants for sunlight.

You may decide that with either method the extra heat stored isn't worth the trade-off. In the case of my small greenhouse, I didn't think it was. I had room for just 24 cubic feet of water storage on the north wall. A rough rule of thumb says that for my Massachusetts climate I should have about 3/4 cubic foot of water storage for every square foot of south-facing glazing. In my case that would mean that I need about 45 cubic feet of water storage. That doesn't mean the 24 cubic feet wouldn't help, but it just didn't seem worth the bother or loss of growing space. This is especially true when the crops grown can stand temperatures down to freezing. It has been rare, indeed, that a supplementary, small electric heater has been needed to keep this greenhouse from dropping below the freezing point. A good shutter system may give far more significant savings with no loss of light and less bother than a heat storage system.

## Ventilation

While you don't want cold winds entering your greenhouse (many people use a double entrance way to the outside for this reason), you do need fresh air passing through to provide plants with carbon dioxide. Moving air is also helpful in preventing disease. Finally, unless you have some sort of storage system, you'll have to vent excess heat even in the winter; with a heating system, you'll have a build-up of excess heat in other seasons.

Probably the simplest solution to the ventilation problem is an exhaust fan controlled by a thermostat. It should be of a size that gives you a complete change of greenhouse air every minute or two. When combined with a louvered opening, it can get rid of excess heat very efficiently. The basic idea is to let cool air in down low at one end of the greenhouse and exhaust the air up high (where hot air gathers) at the other end. This system requires electricity, an extra expense, and is subject to electrical blackouts which could cause the greenhouse to overheat and the crops to die.

There are several small, solar-powered units now on the market that will open ventilators automatically when the temperature reaches a certain point. They use a heat-sensitive fluid in combination with a spring mechanism to open and close vents. Such units are available from Dalen Products, Inc., 201 Sherlake Drive, Knoxville, TN 37922, and Bramen Co., Inc., Box 70-D, Salem, MA 01970. They sell for between $25 and $50 and have varying abilities in terms of lifting power and size of opening.

Large manual vents are also useful, especially in the summertime, but don't overdo it; remember that every opening presents problems of air leakage.

# Growing In the Pit

Even a simple, small pit without elaborate insulation can be an amazing season extender, keeping crops such as tomatoes going months after they've faltered outdoors and providing a good place to get seedlings off to a healthy start weeks before they can be planted outside. But even a well-constructed greenhouse can't manufacture sunlight. The simple facts of life for plants are they need both heat and light, and in the wintertime both are scarce. The sun is low on the horizon and the days are short. In practical terms this means that during the shortest days in December and January, very little growth takes place in any greenhouse. Instead, plants tend to go into a holding pattern, merely maintaining what they have achieved earlier.

### HOLDING ITS HAND

The pit greenhouse requires a lot of hand-holding to keep the temperature warm enough for the plants but, on sunny days even in winter, not too warm for them. So does this mean you can never leave it for a day or so? Nope. And the way to do it is simple. Leave the insulating panels on the south side in place all day. The plants will simply act as if it were an exceptionally long night, and not suffer a bit.

Use this trick, too, when winter days are cold and sunless. If there's no heat to be gained, keep the glazing covered. But remember, only the slightest bit of sunshine will provide a lot of heat for the pit.

## Growth Rate

The net result is you can't look at a seed package and learn, for example, that a variety of lettuce matures in 63 days, and then expect that performance for lettuce planted in November. Nor can you plant tomatoes in September and expect fruit all winter. Chances are the tomatoes will not set fruit before the light gets too low, so plant your greenhouse crop of tomatoes in July.

This is not a message of despair. It just means that if you want to grow vegetables and flowers in your greenhouse all year you'll have to learn new schedules for them. You'll also have to experiment and be on the lookout for special varieties that have a proven track record under glass.

Don't let the relatively low temperatures of the solar pit fool you. Many greenhouse plants, including some orchids, thrive with minimum nighttime temperatures in the 45- to 50-degree range. In fact, some plants need these kinds of nighttime temperatures to perform. If you want cacti to flower, for example, it's a good idea to give them cool nights (down to 40 degrees) and a six- to eight-week dry spell in the winter. These conditions are very easy to achieve in the solar pit. (You'll also want to keep the day temperatures from getting excessive and spurring new growth.)

## Soil Needs

Some growers use sterilized soil. Others prefer pasteurized soil where only the most harmful organisms have been killed, and still others say there is no need for either in a well-balanced greenhouse. Although in other greenhouses I frequently use pasteurized soil, or even a sterile, soilless mix for seed starting and for potted plants, I have never treated the soil in my pit in any special way. Undoubtedly I have run into more weed problems in the pit greenhouse because of this, but so far have encountered no serious disease problems.

I have had especially good luck with lettuce, radishes, broccoli, chives, onions, parsley, tomatoes and peppers in my pit greenhouse, but that far from exhausts the list. I've also used it for seed sprouting, hardening off seedlings started indoors, growing some cut flowers (gloriosa daisies and snapdragons), forcing spring bulbs and maintaining hardy bonsai. Others have found eggplant, Brussels sprouts, cabbage, Chinese cabbage, kale, kohlrabi, spinach, leeks and even beans, to be good solar greenhouse crops. Obviously you want to stay away from large plants that need lots of room, such as corn.

There's still plenty of room for experimentation with crops in solar pits where temperatures are allowed to drop relatively low. If you use some artificial heat to maintain temperatures of 55° or 60° F. (nighttime minimums), then the list of possible plants is enormous and no different than for any other greenhouse.

One hint. If you're using no artificial heat and feel the temperatures will drop below freezing on some nights in your greenhouse, don't give up. The key to holding many plants under these conditions is to water very sparingly and make sure they go into the night dry, rather than having their tissue plumped up with water that will freeze and burst the plant cells.

**BLANKET OF LEAVES**

If frost threatens tender seedlings, either in the greenhouse or outside, a light sprinkling of leaves over the entire scene of operations will do wonders in the way of protection. If the leaves get in the way of growth of the tender seedlings, brush them to one side in the daytime and keep them handy to be replaced on the next cold night.

Helen and Scott Nearing
*Building and Using Our Sun-Heated Greenhouse*
Garden Way Publishing $7.95

# Cost

At 1980 prices, a 8′ × 12′ solar pit, with insulated, concrete block foundation would cost about $500. However, that figure may vary significantly depending on your choice of glazing materials, type of wood, and amount of insulation. (It is assumed all the work will be done by the owner.)

For more details about greenhouse structure and horticulture, take a look at some of the following books. You can use this knowledge more effectively now that you understand the special qualities of a solar pit. Here are some books which I've found especially helpful.

*Building and Using Our Sun-Heated Greenhouse,* Helen and Scott Nearing, 1977, $6.95; available directly from: Garden Way Publishing, Charlotte Vt. 05445. Please add 60¢ postage for orders less than $10.00.

*The Complete Greenhouse Book,* Peter Clegg and Derry Watkins, 1978, Garden Way Publishing, $9.95; available directly from: Charlotte, Vt. 05445. Please add 60¢ postage for orders less than $10.00.

*Winter Flowers in Greenhouse and Sun-Heated Pit,* Kathryn S. Taylor and Edith W. Gregg, available from: Charles Scribner's Sons, N.Y. rev. ed., 1969.